Socks

**By Chuck Baker
With Story and
Illustrations
by Paul Pruitt**

**For Katie
for Christmas 2003**

Newspaper Announcing Pastor Willing's Feat

SOCKS

*T*he Houston Graduate School of Theology had never before had such a famous student graduate from their college. This was none other than Barnaby Willing, astronaut extraordinaire! His face was plastered all over the papers for his *latest flight mission to Mars. There were even a pair of sneakers named after him—"Willing's Wings"— and he had a size twelve pair of them in every color. But this story is not about Pastor Willing and his studies. Nor is it about a flight to a faraway planet. It's about a pair of socks. Space socks, to be specific . . .*

It was tough leading a double life. Here he was, studying to be a pastor, taking exams, handing in the papers, studying Greek, Old Testament, New Testament—and all the while, carrying out missions for NASA. *It was enough to make an organized person go crazy*, he thought. *Let alone a scatterbrain like me.*

But he'd done it, hadn't he? He'd managed to pass his exams, prepare for the mission, and complete both without a hitch. His home looked like a hurricane had hit it, clothes draped over the furniture, every flat surface covered with papers or open textbooks, but he'd done it. Now all he had left to do was clean up. He'd started with the clothes . . .

THE CLOTHES! He'd completely forgotten about them. He'd accidentally sent the space socks with the rest of the clothes to the dry cleaner...the $100 000 space socks! The socks looked like ordinary blue socks but have special quick drying feature which quickly soaks up the sweat when the astronauts are on space walks in the sun and special insulating properties that keep the astronauts very

7

warm on the space walks when the astronauts are in the shade. Temperature differences between sun and shade in space are astronomical.

Don't panic, he told himself. Everything's ok. It's likely that no one knows what they're worth. Still, he rushed to the bedroom to check out the pile of clothes the maid had left folded neatly on the bed.

Pants pressed and neatly folded, shirt collars starched hard the way he liked them, the usual piles of undershirts and underwear and socks—but NO space socks! He searched for the NASA logo on the pile that was there, but the space socks were nowhere to be seen. He couldn't afford this mistake.

Not now. What would NASA think? If he could

misplace a valuable pair of socks, how could they

trust him with a billion-dollar machine? He had to

get them back!

The maid! He'd phone the maid, she'd know

what happened to them, she was the one who took

the clothes to the cleaners and brought them back.

Maybe even the dry cleaner had wanted to send

them back all clean already but since they were

special socks, they were having a hard time

cleaning them. Whatever was used to give them

their quick-drying ability was sure expensive

enough. Maybe it was hard to clean too. *Now

where did I put that number?*, he thought.

He overturned books and chairs and tables searching frantically for the slip of paper that had his maid's phone number on it. Papers were flooding the floor, and the room was even more of a disaster than it had been before. He was tugging at his hair, trying to remember where he put the paper. His hair stuck out and made him look like Albert Einstein on a bad hair day. And then he looked up.

Albert Einstein on a Bad Hair Day

There, where it should have been, tacked up under a magnet on the fridge, was the slip of paper. He dialed the number quickly. "Pick up, please pick up," he said, to no one in particular.

"Hello?"

"Hello, Colleen? Pastor Willing here. I have a huge problem. I'm missing some of the clothing that you dropped off at the cleaners."

"Really? What is it you're missing?"

"A pair of socks, they looked like ordinary socks, but they were different…"

"Did they have N-A-S-A on them, by any chance?" she asked.

"Yes! That's them! Did you find them?"

"No, sir, I only found one with the clothes they sent back. I thought it must have been a mistake, and someone's sock got in with your clothes."

"But I'm an astronaut! Who else's socks could they be? Why didn't you call the laundry back right away?" He was almost shouting now.

He could hear Colleen begin to cry. He thought to himself, she's certainly an excellent maid…but scatterbrained!!

"So what did you do with it, throw it away?" This could now pass for shouting and he overturned the nearest garbage can and looked to see if it was there.

"No sir, I just dropped it off with some of my things at the Salvation Army." She said haltingly, choking back the tears

"The Salvation Army, how could you?!! Do you have any idea how…" Pastor Willing stopped in mid-sentence. *Love, joy, peace, patience, kindness, goodness, faithfulness, gentleness, and SELF-CONTROL.* He recited the fruits of the Spirit in his mind a few times, and told himself to calm down.

"Is, is everything ok, sir? Did I do something really bad?" she offered tentatively?

"It's ok, Colleen. It's not the end of the world, I'm sorry for getting mad at you. Let me start again. Can you please let me know which Salvation Army place you dropped the sock off at?"

"Yes sir, I can sir. It's the one over near the Society for Underwater Technology. You know, the guys who do the ocean stuff. Do you do that too, or are you just a space guy?"

"I'm just a space guy. So the one near SUT, got it! I'll drive right there! Thanks for your help, Colleen!"

"You're welcome sir. I'm sorry for giving away your sock. I didn't know how important it was." Colleen was trying to do her best.

The Salvation Army Shield Symbol

****

*A*bel Humphrey sat on the park bench, and lifted up his size twelve shoe to take a look. Disgraceful, he thought. The leather soles had worn right through…and the socks too. These shoes would never last the winter. And with holes in his socks, he'd be sick in no time.

He'd sold shoes and repaired for many many years, but just last month, he'd been fired by the shoe store. Sure, they said he wasn't as fast on the job as he was before, and a few customers were complaining about waiting…but he was the best in the business in this part of town. He knew shoes. He could sell shoes. It was the one thing he could do. It

15

was just a matter of a new manager fresh out of business school who was still a long way from having the experience he needed, to do a good job. This new manager didn't know a diamond when he had one in his hand. Abel was that diamond.

His friends had always joked about his name…"Abel should be your second name," they'd say. "Your first name should be *not*. You're *not able* to do this; you're *not able* to do that. Who can do this job? *Not* Abel." Sure, they'd all had their fun, but he had focused on doing one thing well in his life, and that was selling shoes and he know few better at it.

Now he was replaced by some young whippersnapper. It wasn't fair.

Things didn't stop there either. After a string of bad luck, he owed so much money to the bank. There was the loan for a new car after he crashed his other one last year, the house repairs after the flood, and of course, he hadn't had flood insurance, and now the bank had called in the overdue loans, and he was out on the street. And what was even worse, he was out on the street wearing worn out shoes. It just didn't get any worse than this…Except of course winter was now coming.

He said to himself, he had nothing now to lose. In an instant, he knew what he would do to get those socks and shoes he needed…and deserved by rights.

Broke Man

****

*P*astor Willing rounded the corner and waited patiently at the light. He looked over and saw two young boys teasing a third. He remembered the taunts he'd heard as a child—

"Barnaby's WILLING to do anything. Are you WILLING to kiss that girl over there, Barnaby? Is SHE the one you like?" He felt sorry for the boy, but the light turned green, and he drove on. The boy would survive—he'd get stronger somehow, like Barnaby had. Even now, Barnaby had the feeling that he was not cut out for what he was doing. Astronaut and pastor—always in the public eye— sometimes, he just wished he had a simpler life.

He parked his car in front of the drycleaners. He'd pick the other sock up here first, before he made his way to the Salvation Army store. He opened the door and a bell jingled and jangled.

"Hello. Did you come to pick something up?"

"No…I mean yes. I mean, I only got one sock back instead of two. I'm sorry, it's just that it's a special sock"

"I think I might just know exactly what you're talking about. Did you lose the sock recently?"

"Yes. And it has some lettering on the side—"

"N-A-S-A?" she asked.

"Yes! Oh, so you DO have it. I'm so relieved."

"No. But I remember that sock because it was made from material that I hadn't seen before, but I put it with the rest of the mismatches because I thought it just came from the NASA gift store in town."

"Well, it definitely was not a souvenir…Where do you keep those mismatches?"

"Well, usually, in the back, in that big cardboard box right there." She pointed to a large cardboard box that looked like it once held a huge TV.

"What do you mean, USUALLY?" the Pastor asked.

"Oh, we just sent a shipment over to the Salvation Army."

Unbelievable! He should have gone there first after all! He rushed out the door without even a goodbye, and made his way over to the Salvation Army Shop. He ran to the door, and went to turn the knob. Locked! He glanced at the sign. It said: *"New Hours Mon-Fri 10 to 5"*

Salvation Army Store Hours Sign

He remembered today was Friday and he

looked at his watch, it was 5:10. He heard a car out

back start up and drive away. Well, that was that.

He'd have to wait until tomorrow. Unless…the

streets were quite empty already; he could probably

find a way in and search for his sock. *What was he*

thinking? A pastor? Breaking and entering?

Wouldn't that make a great headline! He trudged

off to his car, slumped behind the wheel, and drove

slowly home. He'd make sure he was there before

the store opened tomorrow. He had to be the first

customer. He just had to.

****

*I*t is now 9:30 the same night. Abel knew that

this street was always deserted by this time.

A row of shophouses, and no homes. No people out

for walks, no one to see him slip the cardboard into

the door frame and pull down sharply, opening the

lock. If a police officer spotted him, he'd think he

23

was crazy, breaking into a Salvation Army store. Who would break into a place like that?

But Abel knew what he was doing. A place like that didn't have a lot of money to throw around. No security system wired into the police station, no fancy alarm that would be set off the minute he stepped inside. And one thing they would have, socks and shoes, and plenty of them!

He wasn't a criminal. But he did have to stay warm, didn't he? He got in and made his way to the back of the store where he saw piles of unsorted clothes. They'd just been dumped there— there'd been no time for the workers to sort and fold and display them. He tossed clothing this way and that, searching for socks and shoes that might fit his huge feet. He found matches for a few pairs of socks and stuffed them into his pockets—

And then he saw it. Could it be? He pulled the sock from the heap and examined the lettering on the side: N-A-S-A. It was! He had no idea how this sock had gotten here, but he knew all about this sock. The latest issue of Shoe *News* had called them the most expensive socks in the world. And now he had one in his hand. Of course one by itself was worthless, but if he could find the matching one—

****

*P*astor Willing thought he'd get an early night's sleep so he could get up bright and early the next day. As it turned out, he didn't sleep much. He had strange, horrible dreams. He was flying in outer space, getting closer and closer to his target, when all of a sudden, he felt himself being pulled in another direction—sucked into a black hole of some kind. And as he got closer, he could read something at the edge of the black hole: N-A-S-A—he was being sucked into a giant NASA sock!

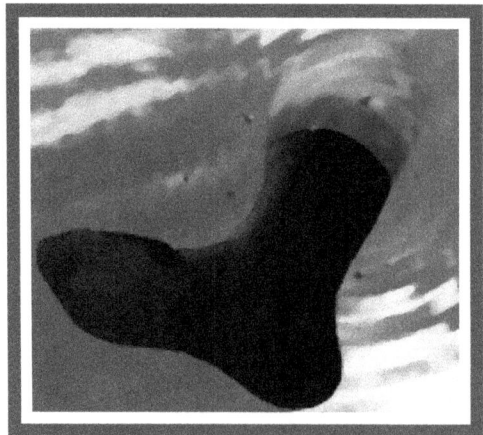

Sock Black Hole

He woke up, sat up in bed, shook his head a few times, then put on his slippers and made his way to the bathroom. He looked into the mirror. His eyes had deep bags under them, and his hair stuck out to the side. He looked terrible—he looked like some of the homeless people he'd seen on the street. Poor fellows. Now that he was a pastor, maybe he should get more involved; try to help the homeless more. Even if he could change a single life for the better, it would all be worth it.

But for now, he splashed some water on his face, put on a baseball cap to hide the bad hair, grabbed a jacket and his jeans, and made his way out to the car again. *He wasn't going to break in, he*

27

told himself. He was just going to make sure his

socks were safely there inside.

****

Rolls Royce

*T*he Rolls Royce of socks. Abel stretched it out and held it up to the moonlight. The lettering was authentic. It really was *the* sock! How could it be here? Well, this *was* Houston after all. Stranger things have happened. . . but this, *this* was like winning the lottery. And now he just had to find the other half of the winning ticket—the sock that matched this one.

From the mountain of clothing in front of him, it looked like he could be here awhile. And he'd have to go slowly. It might be inside a pantleg somewhere, or stuck inside a sleeve. He'd have to check carefully, thoroughly, leaving nothing unchecked.

A mistake could cost him hours of work, hours that he could not afford, because it would be getting light sooner rather than later. He sat to one side of the pile, and slowly lifted up a pair of suit pants.

****

This is turning out to be the longest night of my life, thought Pastor Willing. *First, the missing sock, then, not being able to sleep, and now—I'm thinking of breaking and entering to find it—if it's even there. If they're both there—what are the chances?*

It *had* to be there. It just had to. The stress, the pressure, the worry of messing up for all to see—it was really too much to bear sometimes. A simpler life. All he really wanted was a simpler life. If he could just find these socks, he promised himself, he'd make a change. No more being in the public eye so much any longer. He'd back off for awhile, until he could figure out what he wanted to do with his life.

A simpler life. There had to be a simpler life waiting for him somewhere.

****

*A*rgyle socks, knee-high socks, dress socks, socks with holes in the heel, sports socks, tube socks, rainbow socks with individual toes, like gloves for the feet—there were socks of every kind and shape and size in here, and he was looking for one that was different from all the rest. It was like searching for a needle in a haystack. Nearly impossible, he knew. But he'd be careful. He had

31

time. The store didn't open until 9 a.m. As long as he was gone about two hours before opening, he'd be ok. He had to find that sock!

****

*H*e had to find that sock. He just had to. He had cruised around the block half a dozen times now, without stopping. Each time, he had slowed down near the Salvation Army store, but had kept going. This time, he stopped.

Pastor Willing got out of the car, and slowly walked up to the door. He pulled an old video rental card out of his wallet. He'd watched enough detective movies to know how to do this. In his

heart, he wasn't sure whether he wanted it to work
or not.

****

*A*bel was searching a little more frantically.
He'd seen car lights several times now in
the window, and he was getting nervous. He threw
clothing far to one side now, adding to the pile of
what he'd already seen, what he'd already been
through.

And then he spotted it!

He lunged at the spot where he saw the sock
sticking out—a perfect match!   N-A-S-A ! He
knew he was holding something like $50 000 in his
hands now if he remembered right, which made him

think for a moment that what he was doing had suddenly become more serious. Before, he was just getting warm socks and shoes. Now, he was walking out of here with a lot of money. He never got much of a chance to think about it that night, though.

The next sound he heard was a police siren.

Police Car in Front of the Salvation Army Store

"*H*ey, buddy, we've been watching you. Just what do you think you're do…"

The police officer stopped talking when he saw another person inside the shop running for the back door. He reached for his radio "Dan, go around back, there's another one." The squad car swung around the back of the shop, but it was too late. Abel had already ducked out the back door and into an alley.

Now, all that was left to do was to go question the other guy. At least they had one of the two.

****

"*L*ook, I didn't mean any harm. I'll tell you the whole story," Pastor Willing said frantically.

"Buddy, we watched you come by this shop several times. Your story had better be fantastic. Come with us to the station."

Pastor Willing explained all about the socks. He explained all about just getting back what belonged to him. He was sweating and getting more and more nervous by the minute. He told them about the special socks, but when they heard about socks worth $100 000, they just laughed and gave strange looks to each other. They thought he was crazy!

Dear God, the Pastor prayed, *if ever I needed you, it is right now*.

****

*A*t that moment, Abel walked through the doorway of the police station. He was there at the front of the station, turning himself in, he said. The admitting officer, Laney Edwards looked at him and his clothes. He was obviously a street bum who got too cold and needed a place to stay.

"I'll show you where there's a place that you can get a bowl of soup and a bed for the night," she said.

"No, no, no. You don't get it," he said. "I came in to tell you about the socks. I'm not a criminal, but I do want to report some missing property that has been found. I have a pair of NASA socks worth something like $50,000. I want you to help me find the owner."

Officer Edwards

They looked like a pair of ordinary socks bought from a NASA gift shop. The woman did not believe the man's story one bit. She was about to toss him out, when two officers walked by with Pastor Willing.

"The socks!" he shouted. "Those are the socks I was telling you about!"

"You mean they're yours?" Abel asked.

"You see ma'am? I was telling you the truth."

"Now you two just calm down," she said. "It's time for us to find out the truth."

While both men waited, she made a few calls. Ms. Edwards had a few problems finding the people she needed to talk to, so she sent for the police chief. When Chief Virginia Watson picked up the phone, she soon had someone on the line who could answer Ms. Edwards's questions.

"Yes. Uh-huh. Ok."

"So, what'd they say?" the Officer Edwards asked. "It's just a story, right?"

"No, actually, they're real. But they're not worth $50 000."

"I knew it. They're just a souvenir shop item."

"They're worth over $100 000."

"$100 000 for a pair of socks? That's unbelievable!"

"Unbelievable, but true." Exclaimed the chief.

$100,000 in Cash or at Least a Lot of It

****

*P*astor Willing and Abel Humphrey did their explaining together. What the police finally decided was that since the Pastor had not actually entered the building when they had stopped him, no crime had been committed . . .yet.

And as for Abel Humphrey, he had been honest enough to turn himself in and return the property—even though he was homeless.

Both were free to go.

Pastor Willing was so excited and so proud of the homeless man, that he pulled out his checkbook and wrote him a check for $20 000.

"I'm happy to take it," said the man, "but I have an even better plan, if you'll hear me out."

The two men talked long into the night.

Later that month, they were ready for the big day. They opened the front doors of their shop to a waiting crowd. There had been a lot of publicity about the shoe store owned by an astronaut / pastor and a homeless man.

Now, they opened their doors with pride, and gave each customer the widest smile they could. They were both finally proud of their accomplishments and who they were. Willing had his simple life and Abel had a partner who believed in him. Willing continued to work for NASA in

some capacities, but with greatly reduced duties.
He used the pulpit of the humble cobbler for his role
as a pastor and continued in the Lord's work from
there.

Several pairs of "Willing's Wings" were displayed
in the front window of the shop, but the sign above
said it all: *"WILLING AND ABEL: SHOE
REPAIR AND SALES!"*. There were also some
catchy phrases they both came up with, which can
be seen on the next page.

The End

WILLING & ABEL
SHOE SALES & REPAIR

WE WILL HEEL YOU,

WE WILL SAVE YOUR SOLE,

WE WILL EVEN DYE FOR YOU.

Shoestore Sign

www.ingramcontent.com/pod-product-compliance
Lightning Source LLC
Chambersburg PA
CBHW051259170526
45165CB00004B/1775